# THE QUIET SICKNESS

## A PHOTOGRAPHIC CHRONICLE OF HAZARDOUS WORK IN AMERICA

# THE QUIET SICKNESS

## A PHOTOGRAPHIC CHRONICLE OF HAZARDOUS WORK IN AMERICA

## EARL DOTTER

American Industrial Hygiene Association

2700 Prosperity Avenue, Suite 250, Fairfax, Virginia 22031

Library of Congress catalog number: 98-70894

ISBN 0-932627-85-4

American Industrial Hygiene Association
2700 Prosperity Avenue, Suite 250
Fairfax, VA 22031

AIHA Stock No. 302-CC-98

Design by: Chuck Myers, American Labor Education Center


The staff at AIHA for The Quiet Sickness
is:
Faythe Benson, Manager, Publishing
Christopher Howland, Editor, Books

THE QUIET SICKNESS
A Photographic Chronicle of Hazardous Work in America,
is a major traveling exhibition of 120 images selected
from the 150 duotone reproductions in this volume.

# Dedication

The photographs in this book and exhibit are the result of chronicling the pervasive consequences of hazardous work in the United States. Considering the toll in lives lost and shortened by disease, America's workplaces have indeed created a quiet sickness in which the individual tragedies are largely unseen by the public.

It is to the many workers who endure these conditions daily as they contribute to the most productive society in the world that this book is dedicated. I owe additional appreciation to my historic predecessors, who with camera in hand risked their own well-being to document workers, including children in our mines, mills, and tenements. I am indebted to Jacob Riis, Lewis Hine, Dorothea Lange and W. Eugene Smith, whose photographs forever inform us of the wrenching conditions under which men, women, and children labored only decades ago. They bore witness for those who could otherwise have been so easily forgotten; in so doing, they made deplorable working conditions visible. Their images inspired social and economic reforms whose benefits we take for granted today.

Finally, I have turned to my own family throughout all these years for support and love, as my career made lengthy travel and economic uncertainty too frequent. That burden has been made lighter by my wife, Deborah, and by my children, Michael, Peter, Daniel, and Rachel. Deborah shares my commitment to improving the safety and health of working people and has always given me valuable insight and encouragement.

Earl Dotter

# ACKNOWLEDGMENTS

I particularly wish to thank those individuals who have played key roles in launching my "Quiet Sickness" exhibition program and later making this book, based on the exhibition, a reality.

I credit Howard Frumkin, Allen Tullos, and Tal Stanley with encouraging me to bring an earlier version of the exhibit to the Rollins School of Public Health of Emory University in the winter of 1995–1996. It was Scott Schneider, representing the American Industrial Hygiene Association (AIHA) Social Concerns Committee, who was largely responsible for arranging the first showing of the fully realized "Quiet Sickness" exhibit at AIHA's annual conference in the spring of 1996. And Faythe Benson, AIHA's manager of publishing arranged for the publication of this volume after overwhelmingly positive response to the exhibit at the conference.

I owe a special debt to David Christiani and Ann Backus for bringing the exhibit to The Harvard School of Public Health as part of the school's 75th Anniversary Year Celebration in the spring of 1997 and for arranging a tour of the exhibit to universities, colleges, and private galleries over the next three years throughout the six New England states; to Moe Foner, the originator of the only cultural arts program within the labor movement, for bringing the exhibit to Gallery 1199 in the headquarters of the National Health and Human Services Employees Union in Manhattan, and for showcasing virtually all my exhibits since the 1970s; and Jim Albers, for the highly successful presentation of the exhibit at the main branch of the Cincinnati Public Library in the late summer of 1997, sponsored by the Cincinnati Labor History Society.

There are many individuals who have supported my efforts throughout my photographic career to whom I will always be grateful: Rob Amberg, James August, Cornell Capa, Robert Coles, Mimi Conway, John Dewey, Paul Elfenbein, Dave Elsila, Alex Harris, Edith Holbrook, Michael Holland, Barbara Jenkins, Susan King, Charles Levenstein, Barry Levy, Ann Mack, Davitt McAteer, Jim Melius, Chuck Myers, Karen Ohmans, Bill Ravanesi, Naomi and Walter Rosenblum, Cecil Roberts, Bill Serrin, Joel Shufro, Len Speier, Don Stillman, Helen Stern, Richard Trumka, Gregory Wagner, Dave Wegman, Jim Weeks, and Matt Witt.

# Contents

# Introduction

Scott Schneider, CIH
AIHA Social Concerns Committee

Industrial hygiene is so much more than hanging pumps and taking samples. Our primary objective is to protect the health of working people, but the technical and scientific aspects of our job sometimes divert our attention away from the goal of all this information gathering — to improve the quality of working life for those laboring in adverse conditions.

With that in mind, the AIHA Social Concerns Committee arranged a photography exhibit by Earl Dotter at the 1996 American Industrial Hygiene Conference & Exposition (AIHCE) in Washington, D.C., to remind visitors that real people depend on us to help eliminate job-related fatalities, injuries, illness, and disease. The AIHA Social Concerns Committee has strived for more than 20 years with the purpose of helping members focus on the social impact of their work. Every day, industrial hygienists and occupational health professionals face ethical dilemmas in which they have to weigh the consequences of their decisions and choose what methods will best protect workers. In making these choices they have to consider not just the numbers but the faces and lives of real people.

I firmly believe that the AIHCE exhibit (titled "The Quiet Sickness: Occupational Hazards in the United States") enabled our members, and others, to comprehend this sentiment quite well. More than 120 photographs were featured, and hundreds of attendees walked away clearly moved by Earl's work. After purchasing one of Earl's photos of a coal miner, in fact, one person who worked for a coal company said to me, "I want to hang this photo in my office to remind myself every day why I am doing this job."

These unsolicited testimonials are a tribute to the power of Earl's work and the reason we asked him to put his photographs on display.

Earl's work is vivid and insightful. As he describes, "When I walk through a mine, mill, or factory, I find myself drawn to those subjects who emanate a sense of personal worth and belonging to the human family. When I experience tragedy in the workplace — death and disability — I use the camera to explore not just the person or event, but my own reaction to it. If I am successful, then the viewer will be better able to stand before the photograph and feel the intensity of the moment as I myself felt it."

Undeniably, the impact of some of Earl's photos is enormous. In 1980, when his photograph of a worker wasting away from Brown Lung disease appeared on the front cover of a cotton dust standard brochure issued by OSHA, one of the first actions by Reagan appointees at the U.S. Department of Labor was to order them destroyed because they "lacked objectivity."

Images such as these can and do make a difference and move people.

After the 1996 AIHCE exhibit closed, it was clear there was a need to publish these photos in a book to make them more widely available to the occupational health community and others who have come to know Earl's work through its use in textbooks, health and safety manuals, national magazines and newspapers (including the *New York Times*), union publications, and by OSHA and NIOSH. This book is the result of that effort. We in the AIHA Social Concerns Committee hope this landmark achievement puts a human face on those American workers confronted every day with occupational safety and health hazards — hazards that demand our attention. In doing so, occupational health and industrial hygiene will extend beyond the realm of mere statistics. The focus can once again be on the real lives of the people we are trying to protect.

# Foreword

Robert Coles
Editor, *DoubleTake* magazine
Center for Documentary Studies at Duke University

As I look at Earl Dotter's photographs, I hear the voice of a West Virginia miner I got to know in the late 1960s while working in Appalachia, trying to learn from children what it means to live up a hollow of, say, eastern Kentucky or in one of the company towns built by "Big Coal" in the early 20th century. The miner's children were among those I had come to know, and he often asked me how I thought they were "doing" — how I judged their elementary school education to be "coming along."

Once, after yet another discussion of classroom life as his two sons and two daughters were experiencing it, he spoke of himself, his own all too brief experience with formal education, and his prospects: "I tell my children their big hope is to learn their letters and their numbers so they can go live someplace else — in Beckley [West Virginia] maybe, where my younger brother lives, and he's become an accountant. Here you go down to cut the coal, or you get what you can from the county folks [welfare payments, food stamps]. If you become a miner, it don't take long before you're coughing and you can feel it eating your lungs away, and your spit is dirty with the [coal] dust, and you know what's ahead, you sure do: trouble breathing and more trouble breathing, and down the road, you won't be going down the mine; you'll be down all right, flat on your back, and that's how it is. There's no turning back, once you're in this digging life — you're racing against the day when your chest says: I can't take it no more; I'm just all messed up with that coal dust — it's just done me in for good."

I remember presenting his words in conjunction with his pulmonary X-rays to a group of my medical colleagues — a way of giving some "flesh," as it were, to a discussion of "Black Lung disease." I remember the consternation of several doctors. Why do such men keep at it, remain miners, they intimated. Why don't they quit rather than submit to what is so often a slow death?

It is, in that regard, hard for any of us to put ourselves in the shoes of others — to understand, in this case, the mix of desperation and pride that informs the life of an extremely hard-working coal miner who wants so very much to be a bread winner for his family, at all costs, even his life. For many of us this life is gener-ous. We pick and choose our way through various possibilities around various corners. For that man, for many thousands like him, there is the felt, fateful finality of either/or: one takes the job available, or experi-ences the futility, the despair, the sense of worthlessness that goes with unemployment.

No question if I had brought some of my taped interviews with miners to our discussion so many years ago, I would have been able to give my medical colleagues a better sense of those "Black Lung patients" (as they were repeatedly called that day). I suspect, too, that we all would have spoken more sensitively had Earl Dotter's photographs of Appalachian miners been available to us: those miners' eyes looking at ours; their clothes, their skin, their bodily postures right there for us to see, to contemplate — fellow citi-zens, but in so many ways vastly apart from us. For many years now, Earl Dotter has enabled us to see these individuals, those who work in mines and factories and fields and offices and buildings as they are being built or have been built — men and women whose labor enables the rest of us to keep warm or eat well or be sheltered, or have a place to work, or drive a car. His camera has wandered the American land, graced by an artist's talent but prompted by a citizen's moral energy — an insistence that those who spend their days doing valuable but hazardous work deserve our alert, worried attention. His camera has made a record of these individuals and their daily toil — the circumstances in which they find themselves as they strive to make a living. His camera has asked us to consider others as well as ourselves — such as the miner quoted above, who once pointed out: "Some of us have no choice but to get hurt slowly so we can bring in a paycheck every week; otherwise, your family gets hurt real fast."

There are, of course, those who will listen to such words, view the photographs that appear in the pages that follow, and shrug their shoulders. That's the way it is, and has to be, they will say. There are always jobs that put people at risk. Yet the story of the industrial West is the story of a gradually increasing public sensitivity to the hurt and pain of what used to be called the "laboring classes," those who give of them-selves mightily so that machines work, goods get produced, and a society grows and flourishes.

What Lewis Hine depicted so many years ago — children doing the hardest work under the most unfavor-able conditions — would now be regarded as outside the pale, an intolerable affront to a nation's dignity, let alone that of any particular person. Not that Hine's photographs, in and of themselves, enabled us to have child labor laws enacted — even as the powerful, brilliantly idiosyncratic *Let Us Now Praise Famous Men* of James Agee and Walker Evans, or the vigorous, federally supported documentary photography of the 1930s (that gave us the still evocative work of Dorothea Lange and Ben Shahn and Marion Post Wolcott and Russell Lee) made a decisive difference in the way migrant farm workers and sharecroppers planted, cultivated, harvested, and lived their quite marginal, vulnerable lives.

Still, politics has to do (among other things) with opinion, and no question, as a consequence of the devotion of certain photographers, several generations of Americans, in substantial numbers, have been brought close to scenes and sights that surely prompt concern, alarm, sadness, and no small measure of moral indignation — a sense of shame that so many have to pay such a stiff personal price (mean living, illness, even death) so that others can do well.

That phrase "moral indignation" is no small matter in a democracy — nor in art, or for that matter, psychology. How well I remember hearing these words from Sen. Robert F. Kennedy as a group of us doctors were about to testify in 1967 about what we had recently seen among the impoverished children of the Mississippi Delta (serious vitamin deficiency diseases, for one, supposedly a thing of the past in mid-20th century America): "Tell us what you saw, make us see what you saw — and the country will be stirred to action."

We tried hard with our words, but pictures, of course, would have helped make our remarks all the more compelling. Some of the photographs in this book don't even need words — a careful look tells the story. This or that misfortune or injustice or wrongdoing or potential danger is spread visually before us, whereupon it is up to the person holding the book to decide what response ought to be forthcoming.

In this regard, when the psychoanalyst Erik H. Erikson was bringing his important book *Childhood and Society* to a conclusion, he emphasized the importance in our mental life of judicious indignation, without which "a cure is straw in the changeable wind of history" — his way of reminding all of us ever so psychologically attentive readers how valuable an expressed righteousness can be, not only for those in need of social and political change, but for the rest of us, who may be quite well off (safe from the dangers facing the people in the pages ahead) yet enmeshed in our own kind of ethical and emotional jeopardy. What happens to us when we become indifferent to the plight of others, when we gradually become all too evidently self-preoccupied?

Put differently, it is in the self-interest of all of us — it is in our national self-interest — that this "chronicle of hazardous work," which unfolds so vividly and tellingly (a testimony to a dedicated and accomplished photographer's unyielding, tenacious persistence), be regarded as a cautionary tale, as a warning that ought not go unheeded. This book is, if you will, a moral witness of sorts, sustained over many years. An observer has traveled far and wide, seen what ails and threatens many of our working people, and made a record of what he has seen. Now it is our turn to absorb the lessons of these pages, to consider what ought to be done. Many of these people belong to Dostoevsky's "insulted and humiliated": they edge closer to injury and worse by the day; they deserve better of this life, and we who get to know them and what threatens them, through this book, have a chance to affirm our own worth by standing up on behalf of their worth.

"In a community of souls, an injury to one is an injury to all," Dorothy Day once remarked as she described a communitarian philosophy to which she had given so much of her life (spent ministering to the poor). So it ought go in our American community, and so one hopes and prays Earl Dotter's photographs will help many of us realize.

*"Art is not a mirror held up to reality but a hammer with which to shape it."*

**—BERTOLT BRECHT**

# HEALTH CARE

Needles found in soiled bed linens shipped from local hospitals to the Brooklyn Central Laundry.
These sharps were found by laundry employees over a one-year period.

*Brooklyn, New York (1997)*

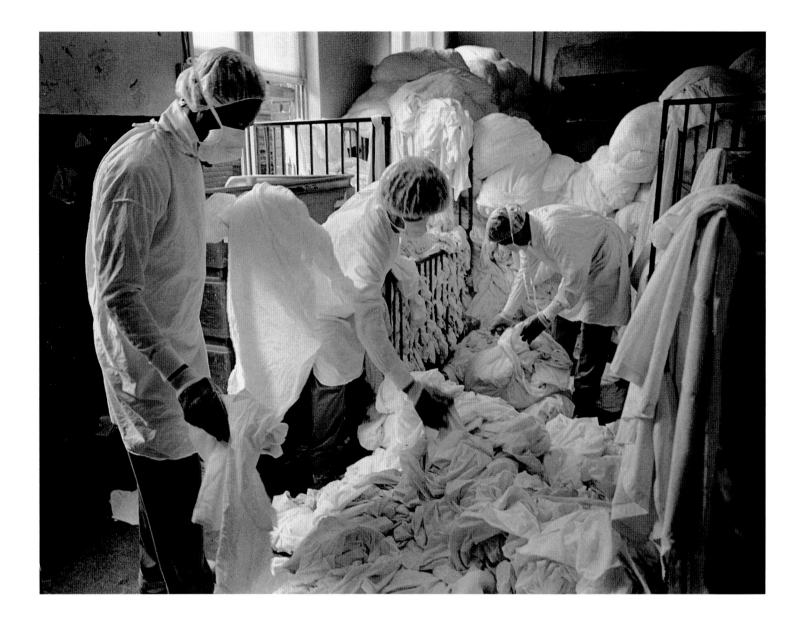

Hospital laundry workers sort soiled bed linens from wards with patients being treated
for contagious and infectious disease. Wire mesh gloves lower the risk of needle sticks.

*Bronx, New York (1982)*

Nurse in intensive care unit attempts to stabilize
a gunshot victim at D.C. General Hospital.
Understaffing compromises patient emergency service
and stresses critical care staff.

*Washington, D.C. (1989)*

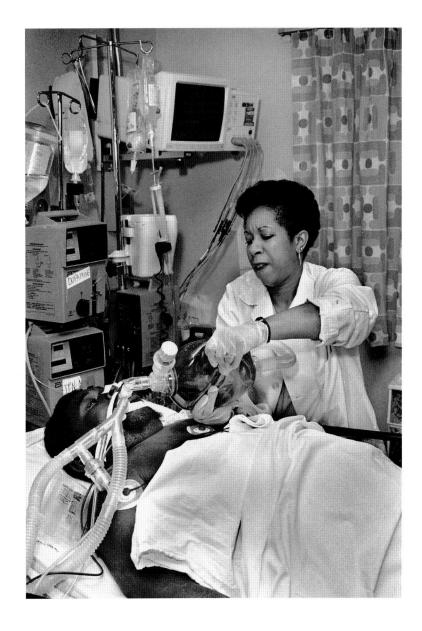

This laundry worker stocks poorly designed hospital
bed linen storage racks, forcing her to lift heavy
bedding above shoulder height.

*Chicago, Illinois (1993)*

A hospital worker prepares to sterilize surgical instruments. The sterilizer uses ethylene oxide, a toxic chemical still in wide use in hospitals today. Exposure to ethylene oxide is associated with severe pulmonary irritation.

*Bronx, New York (1982)*

As the sterilizer is opened to remove surgical instruments, the employee is engulfed in steam laden with ethylene oxide vapor. Within a few seconds after this photo was taken, he was completely obscured from view.

*Bronx, New York (1982)*

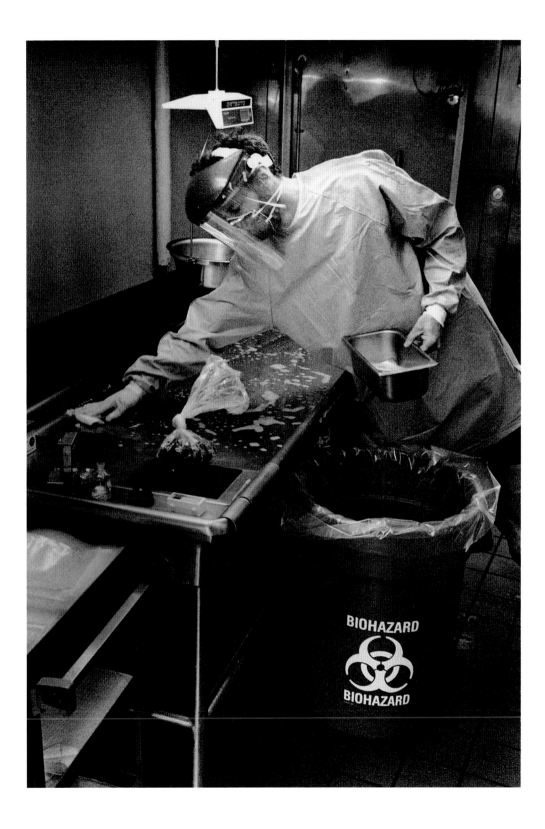

An animal pathologist disposes of the carcass of a dissected rabid raccoon at a state wildlife pathology unit. Pre-immunization for rabies is required just to enter this room.

*Delmar, New York (1997)*

# PUBLIC SAFETY

A slain police officer's son remembers his father at dedication of the National Law Enforcement Officers' Memorial. The National Safety Council reports that public service employees are twice as likely to be killed or seriously injured on the job as other workers.

*Washington, D.C. (1991)*

This maximum security prison corrections officer must work
unarmed among prison inmates.

*Coxsacke, New York (1984)*

A Maine State Prison corrections officer monitors inmate population during mass release. Violent incidents most often occur during this interval while inmates proceed to their daily work assignments.

*Thomaston, Maine (1997)*

Parents of Tammy Sperle mourn the loss of their 34-year-old daughter who was murdered while she worked as a storekeeper in a men's prison in Michigan. In 1996, more than 600 public workers lost their lives on the job and some 550,000 suffered severe injuries.

*Saline, Michigan (1997)*

Volunteer firefighters battle a mobile home blaze.
Toxic smoke from synthetic building materials confront the firemen.

*Buffalo Creek, West Virginia (1976)*

Firefighters await a call to action. Ruptured natural
gas lines ignited a construction trenching operation.

*San Francisco, California (1980)*

# CONSTRUCTION/ MAINTENANCE TRADES

Window washer on the 86th floor of the
Empire State Building with a view toward midtown Manhattan.

*Borough of Manhattan, New York (1986)*

A city bridge inspector secures his tie line on the condemned Strawberry Mansions bridge. More than 8 million public employees in 27 states do not have federally approved OSHA programs.

*Philadelphia, Pennsylvania (1997)*

This bridge inspector, prior to entering a confined space, prepares to check for hazardous air quality in a chamber of the Hood Canal Floating Bridge.

*Near Port Gamble, Washington (1997)*

After completing his inspection, the inspector heads back to the surface.

*Near Port Gamble, Washington (1997)*

A snowplow driver installs chains as he prepares to clear deep snowdrifts on the highway ahead. Operators must be ever-vigilant to avoid collisions with stranded vehicles and motorists obscured in the snowbanks in whiteout conditions or darkness.

*Tracy, Minnesota (1997)*

A residential trash pick-up crew member runs an average of 20 miles every day behind a city garbage truck. Vehicular traffic and repeated lifting while on the run causes thousands of crippling disabilities each year among sanitation crew workers.

*Indianapolis, Indiana (1997)*

This lineman begins restoring power lines after hurricane damage.

*Houston, Texas (1982)*

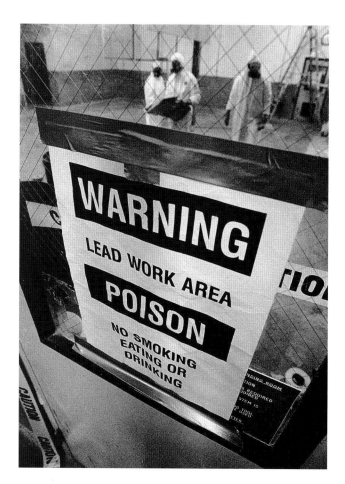

A lead paint abatement area is posted with easily visible warning signs at all entry points. It is isolated from other less hazardous operations. Lead-based paint is being removed from the skin of older commercial aircraft during rehabilitation.

*Everett, Washington (1996)*

The abatement crew dons personal protective gear prior to entering the paint removal area.

*Everett, Washington (1996)*

Left, a demolition worker cuts steel girder with acetylene torch. According to NIOSH, 3491 construction workers fell to their deaths between 1980 and 1989.

Above, a demolition site truck driver signals the bulldozer operator unloading building rubble. One thousand construction workers are killed on the job each year, more fatal injuries than in any other industry.

*Times Square, New York City (1990)*

An asbestos tear-out crew member, wearing personal protective gear, holds bagged asbestos insulation labeled with appropriate warnings.

*Borough of Manhattan, New York (1997)*

Licensed certification IDs of multiethnic asbestos tear-out crew are posted at a warehouse job site. The warning signs are in Spanish, Polish and English.

*Borough of Manhattan, New York (1997)*

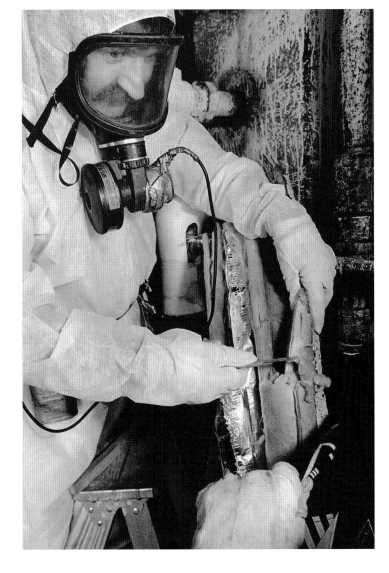

An asbestos tear-out crew member removes insulation from pipes. The hazardous fiber is immediately bagged and sealed for disposal.

*Borough of Manhattan, New York (1997)*

# AGRICULTURE/FOOD PRODUCTION

A 17-year-old girl operates an antiquated family farm tractor with no machinery guards or roll-over protection. Every year, about 300 children are killed and 23,000 children and adolescents are injured in agriculture-related activities, as reported by NIOSH.

*Cambridge, Maryland (1992)*

Twenty-two farmworkers stoop to harvest, wrap, and box iceberg lettuce for shipment direct from the farm field. The speed of the mechanical equipment sets the work pace.

*Watsonville, California (1996)*

This farmworker stoops to harvest a lettuce crop.

*Salinas Valley, California (1994)*

A heat-stressed farmworker takes a water break with temperatures pushing 105 degrees. Skin cancer affects a high proportion of farmworkers exposed to long hours in harsh, direct sunlight.

*Friars Point, Mississippi (1990)*

Farmworkers use self-made gear to protect their skin, hearing, and respiratory systems from chemical pesticide dust and noise while working on a mechanical tomato harvester. Farmworkers are at high risk for fatal and nonfatal injuries, work-related lung diseases, noise-induced hearing loss, skin diseases, and certain cancers associated with chemical use and prolonged sun exposure.

*Mendota, California (1996)*

A farmworker gets ready to dust a vegetable crop with pesticides. His son looks on.

*Salinas Valley, California (1994)*

An apple harvester faces many risks, including pesticide exposure, falls from ladders, and the stresses associated with seasonal, low wage employment. The orchard is posted with a state-mandated warning sign.

*Wapato, Washington (1996)*

A 13-year-old unloads the day's crab catch from the family fishing boat. Children
working in the fishing industry suffer a high rate of injury and fatal accidents.

*Deal Island, Maryland (1990)*

College students repair nets after having signed on as part of the crew on
a commercial fishing vessel. Hoping for short-term financial gain, such
inexperienced crews often face great danger in the Alaskan fishing grounds.

*Seattle, Washington (1996)*

A 16-year-old short order cook is at work in a fast food restaurant. He faces hazards from violence to cuts and burns. Each year, about 70 youth under 18 years of age die from injuries at work, and 64,000 require treatment in hospital emergency rooms.

*Salisbury, Maryland (1990)*

Teenagers employed in a fast food establishment.

*Washington, D.C. (1987)*

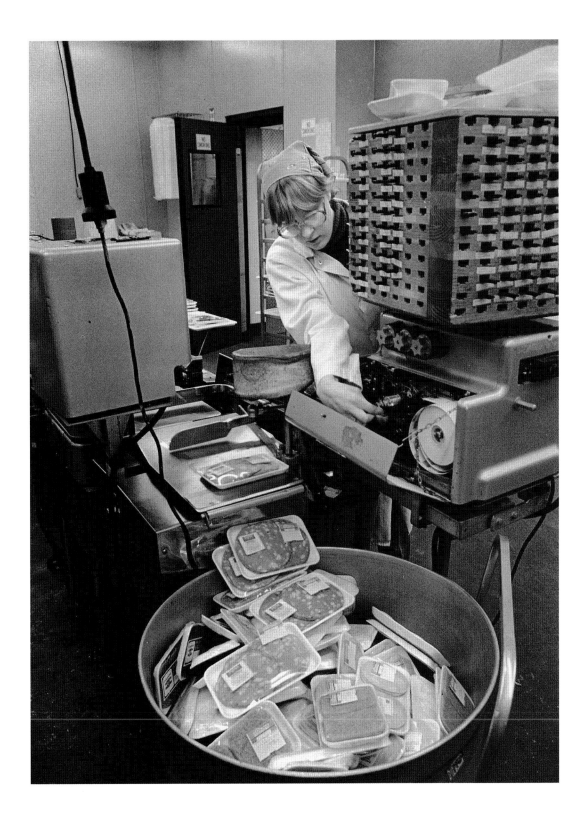

An underage supermarket meat packer unjams labeling machinery. Meat processing machines pose significant hazards to young workers.

*Borough of Queens, New York (1990)*

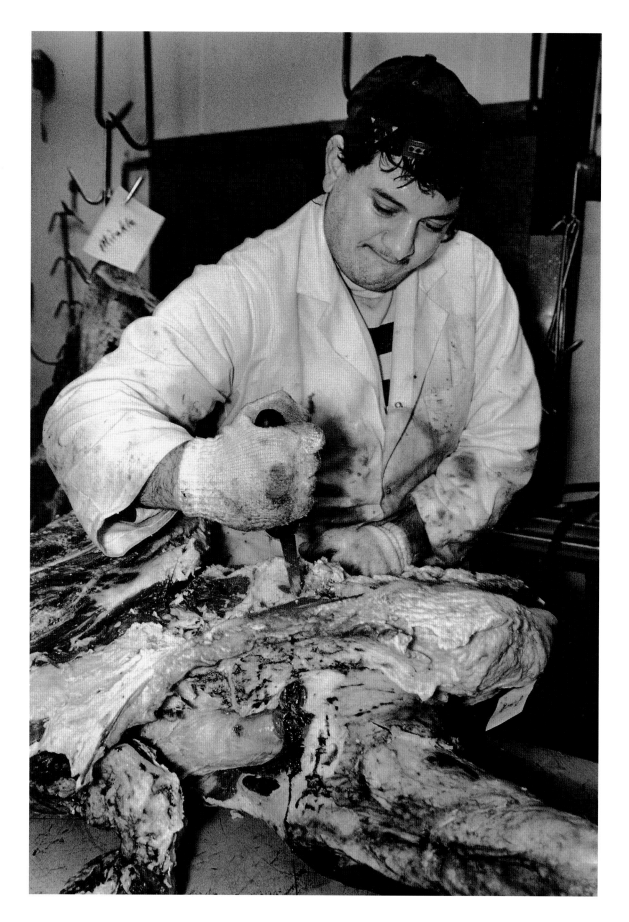

A supermarket meat cutter breaks down a quarter of boxed beef.
He is less prone to repetitive motion injury because his tasks vary.

*Borough of Queens, New York (1985)*

A packing house meat cutter breaks down a side of beef. Meat cutters face multiple hazards,
from repetitive stress to back injuries. Packing house workers handle heavy, awkward cuts of
meat in wet, slippery conditions. Knife cuts are a constant threat to hands and fingers.

*Baltimore, Maryland (1996)*

This poultry farmer uses a respirator in an attempt to protect herself from breathing airborne fecal matter in the chicken house. At right, one of thousands of dead chickens each poultry farmer must dispose of daily. USDA officials estimate individual farmers bury up to 150 pounds of dead chickens per day. The buried carcasses become cesspools of bacteria, leaching into groundwater and, in some cases, local streams.

*Mayfield, Kentucky (1993)*

Shirley Mack is totally disabled with repetitive stress injuries incurred while working one year at a local chicken processing plant. With obvious pain, she holds her daily medications.

*Fayetteville, North Carolina (1997)*

# TEXTILE/GARMENT INDUSTRIES

Tundra Lucas was picking cotton at 10 years of age. Despite mechanization of the cotton harvest, children continued to work in the cotton fields of Mississippi well into the 1980s. Mississippi, in fact, did not enact a compulsory education law until the mid-1980s.

*Bolivar County, Mississippi (1978)*

Children help unload hydraulically operated bin of cotton harvester.

*Bolivar County, Mississippi (1978)*

A cotton buyer's warehouse sign.

*Gilliam, Louisiana (1978)*

Cotton bale opening room workers have the dustiest, most unhealthy jobs in
the cotton mill reserved for them, usually because of discriminatory practices.

*Ware Shoals, South Carolina (1978)*

A cotton mill loom tender must retie threads that break on the loom. When a thread breaks, the loom shuts down, causing the tender to lean well into the dusty machine to fix the broken thread. The tender keeps watch on 30 to 40 dust-laden looms. Working in close proximity to the raw cotton fiber, he has a high risk of contracting byssinosis or Brown Lung disease.

*Ware Shoals, South Carolina (1978)*

Elsie Morrison has just returned home from her shift at the local cotton mill to do child care. Her grandson's mother just went to work at the same mill. Morrison's coughing roused her grandson from a late afternoon nap.

*Erwin, South Carolina (1978)*

This textile mill village home is part of a community of residents with Brown Lung disease. Their jobs at the local mill exposed them to cotton dust. Byssinosis victims usually cannot sleep through the night due to constant coughing brought on by the disease. Community residents had these signs erected to help the victims rest any time of the day.

*Greenville, South Carolina (1978)*

Louis Harrell was the first president of the Carolina Brown Lung Association. This portrait of Harrell was used on the cover of OSHA's Cotton Dust Standard brochure in 1980. Thorne Auchter, in one of his first official acts as OSHA director, destroyed 135,000 copies of the literature. The brochure was revised without photos. Harrell received this plaque from the J.P. Stevens Company when he retired after "28 Years of Loyal and Faithful Service." Two years and one day after retirement, Harrell died at age 62.

*Roanoke Rapids, North Carolina (1978)*

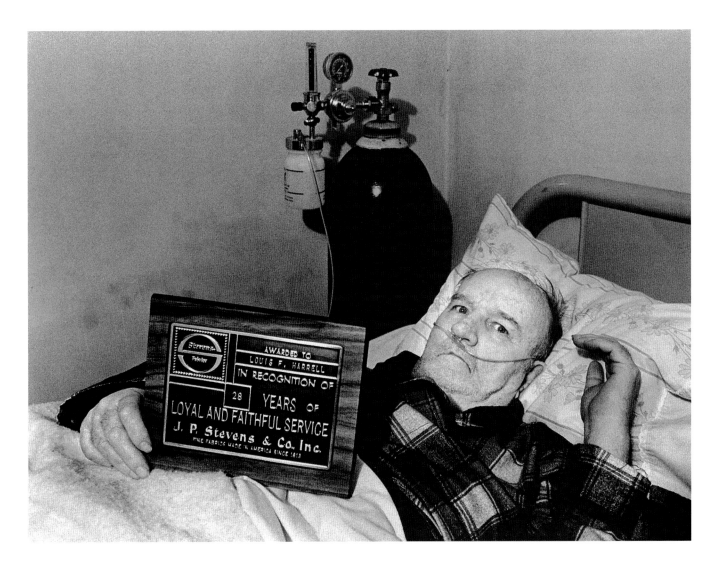

Members of the Carolina Brown Lung Association bid their final farewell to Louis Harrell.

*Roanoke Rapids, North Carolina (1978)*

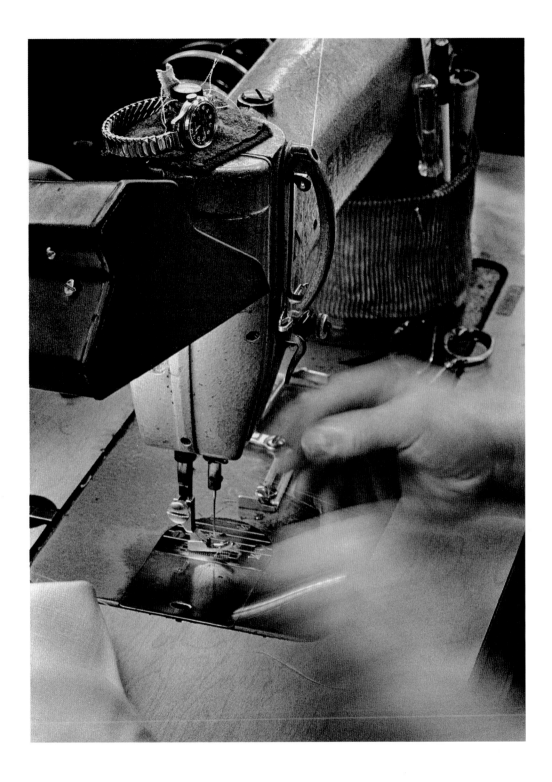

Top left: A child labor task force inspector questions an underage garment worker working during school hours in Manhattan's Fashion District. *(1992)* Bottom left: Garment workers sew bathing suits in crowded, overheated shop. Above: A garment worker times her production level. The "piece rate" system encourages productivity, but often at a cost. Repetitive motion injury shortens the healthy working life of a high percentage of garment industry workers.

*Long Island City, New York (1985)*

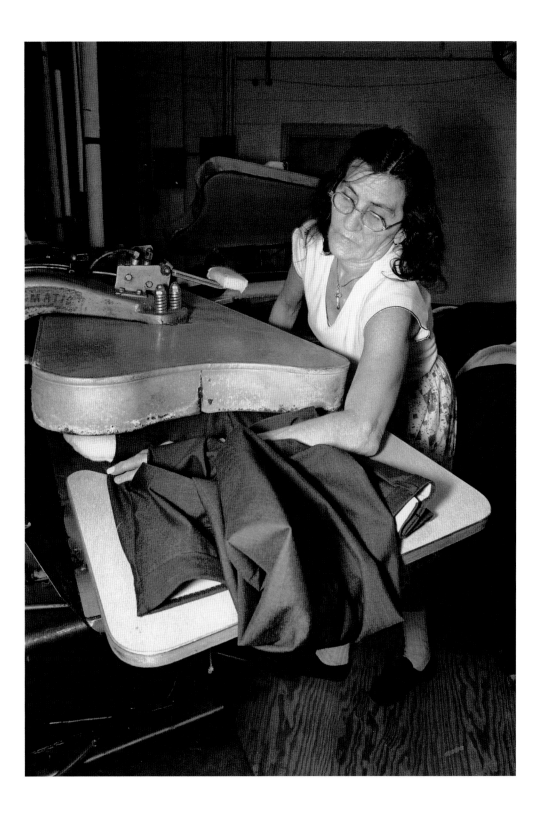

A garment shop presser works on the piece rate production system. In a hurry to maximize her income, she is exposed to burns and heat stress while standing long hours in a crouched position.

*Buffalo, New York (1991)*

This garment worker reacts to the high cost of her prescription drugs.

*Borough of Manhattan, New York (1980)*

# OFFICE/COMMUNICATIONS

Injured hands of a USDA data entry clerk at
work on her computer keyboard.

*Rosslyn, Virginia (1997)*

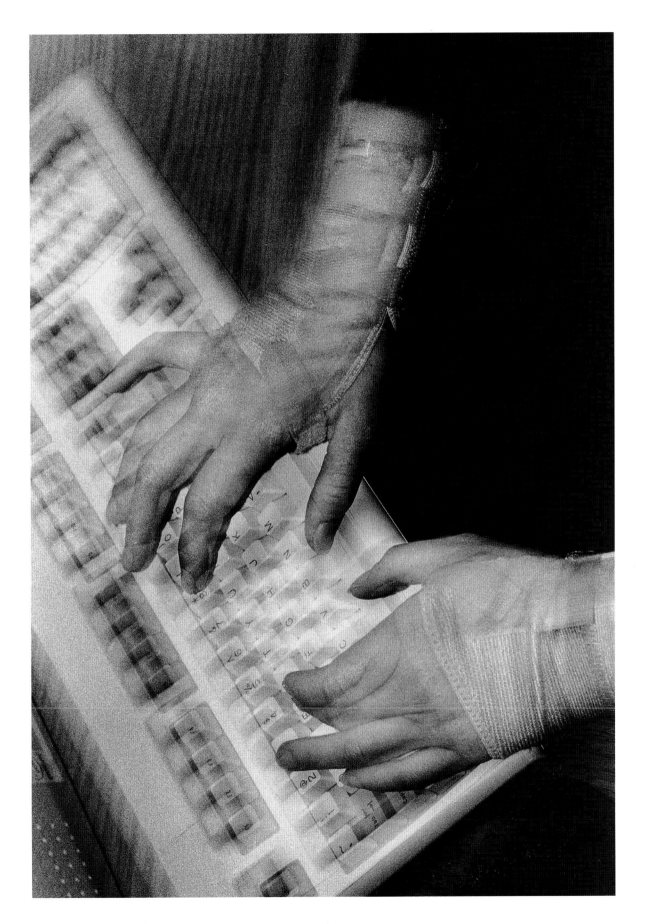

Long distance telephone operators work in relative isolation while
their performance and productivity are constantly monitored.

*Austin, Texas (1980)*

Understaffing of the city tax collection
department overburdens this clerical worker.

*Orange, New Jersey (1980)*

A telephone company technician tests switching
circuits, requiring reaches from overhead to floor level.

*Borough of Manhattan, New York (1981)*

A USDA data entry clerk massages her hands injured over a period of eight years
at a computer keyboard. According to the *New York Times*, repetitive stress injury
reports have increased tenfold in the past 10 years, at a cost of $20 billion to $100
billion a year, depending on how the cost of lost time work is calculated.

*Rosslyn, Virginia (1997)*

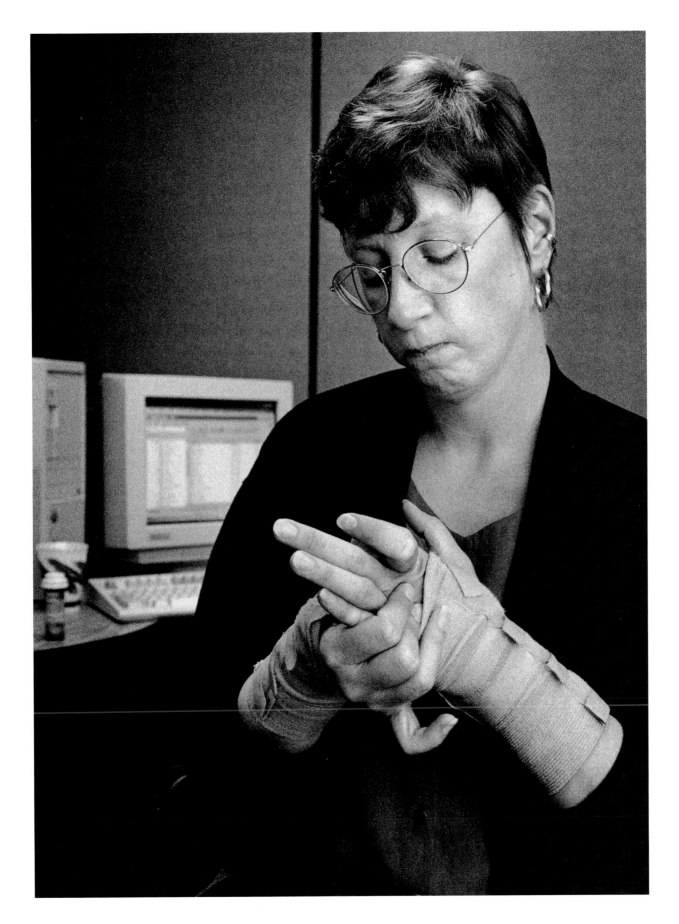

# CHEMICAL/WORKPLACE ENVIRONMENT

This chemical refinery worker takes a benzene sample to monitor product quality. Personal protective gear, inspected regularly for effectiveness, is essential to avoid life-threatening exposure to this carcinogen.

*Westville, New Jersey (1978)*

Offset printing press operators are exposed to solvents through direct skin contact as well as from inhaling the fumes. This press operator wipes the paper feed mechanism with petroleum-based solvent without glove protection.

*Shirlington, Virginia (1988)*

A chemical hazard response team encapsulates chemical vapors with foam to contain a spill during a drill to practice emergency procedures.

*Cincinnati, Ohio (1989)*

A warning sign for chemical workers is provided at access point to benzene storage tank hatches. Benzene exposure can cause leukemia.

*Westville, New Jersey (1978)*

Chemical plant air pollution has an adverse impact on a children's residential playground close by.

*Hopewell, Virginia (1978)*

Chemical lab hazardous waste disposal with appropriate safeguards.

*College Park, Maryland (1991)*

A sign warns local fishermen of dioxin-contaminated fish. The river was polluted with dioxin effluent from the local paper mill in the making of bleached white computer printout paper.

*Pascagoula, Mississippi (1991)*

# COAL MINING

Coal miner Lee Hipshire has just emerged from the mine at the end of the day shift. At age 46, he had worked 26 years underground. A few years later, Lee took early retirement because of Black Lung disease, or pneumoconiosis. He died at 56.

*Logan County, West Virginia (1976)*

A coal miner in a classic posture walks past the cutting bits of a continuous mining machine. Mining has the highest rate for fatal injuries of all U.S. industries, according to NIOSH reports.

*Clearfield County, Pennsylvania (1976)*

Spending a shift in a 30-inch seam low coal mine, this coal cutting machine operator remarked, "It's a bit like working under your kitchen table all day."

*Logan County, West Virginia (1976)*

Buck Koptchak tests the torque on a coal mine roof expansion bolt.
Two years earlier, his father lost his life doing the same job.

*Clearfield County, Pennsylvania (1976)*

This coal miner is setting temporary roof supports, the most dangerous
underground mining job. Roof falls continue to be the leading cause of
injury and death in coal mines.

*Clearfield County, Pennsylvania (1976)*

This longwall coal mining shear operator wears protective head gear,
which also provides filtered dust-free air under his face mask.

*Bentleyville, Pennsylvania (1983)*

A coal mine section crew takes its midshift meal in a specially reinforced area called the dinner hole.
Ordinarily, it is not a safe practice for an entire crew to gather in the same area of an underground mine.

*Near Farmington, West Virginia (1976)*

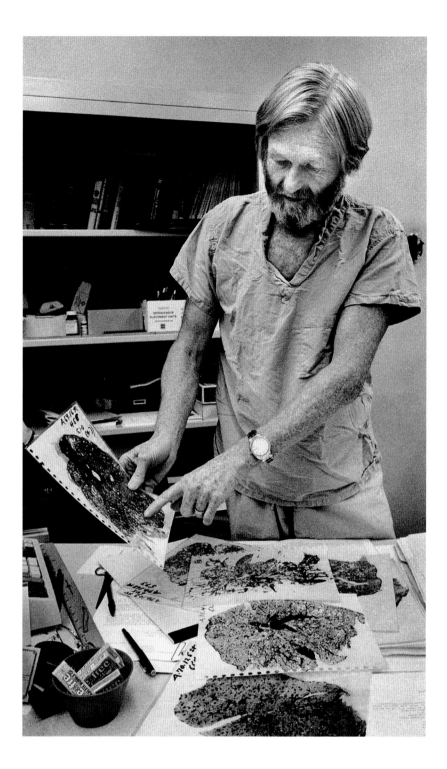

Above: Black Lung doctor Donald Rasmussen examines lung sections from coal miners who died from the disease.

*Beckley, West Virginia (1977)*

Right: A Black Lung victim nears his last breath. His sons, also coal miners, keep a vigil. NIOSH estimates 2000 miners die every year from lung diseases caused by exposure to coal dust.

*Logan County, West Virginia (1976)*

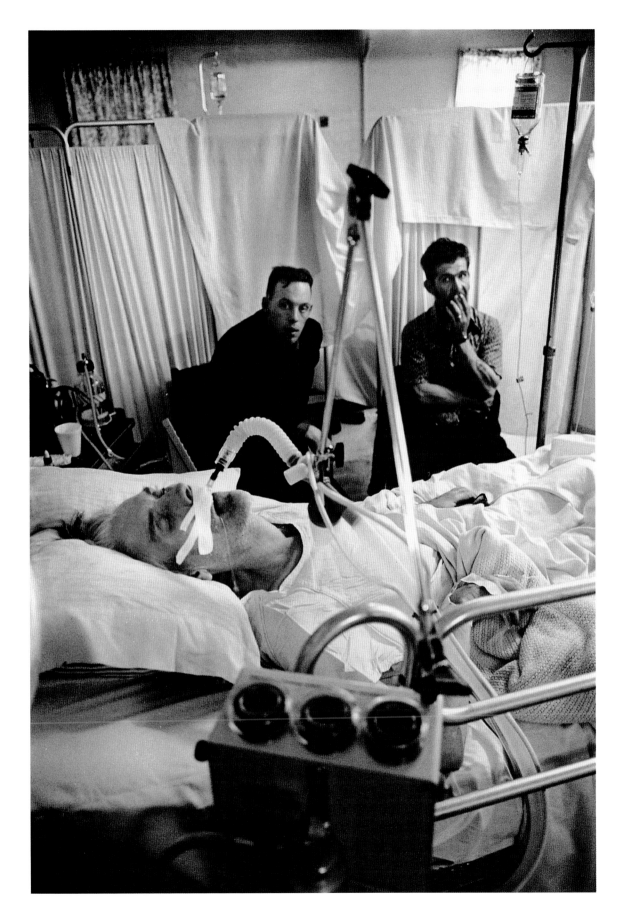

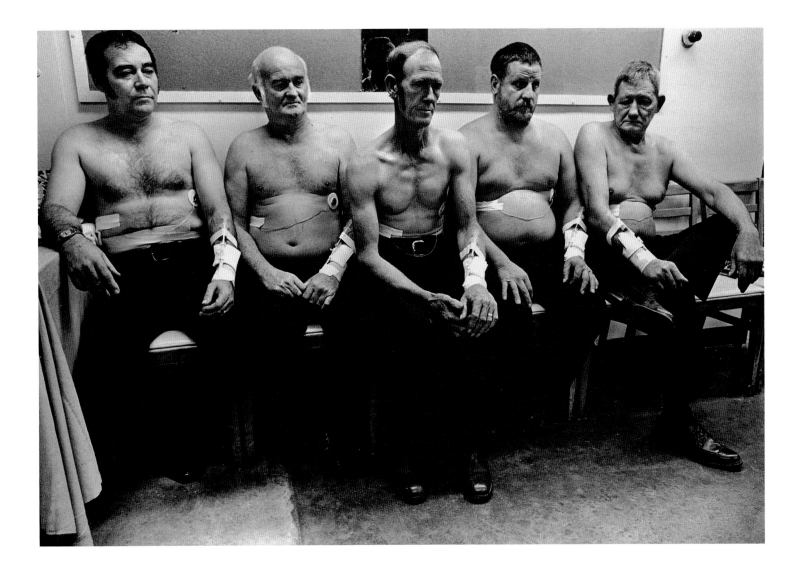

Black Lung victims await testing. An extensive series of X-ray and lung function evaluations are required to determine if federal Black Lung compensation payments will be paid.

*Beckley, West Virginia (1976)*

Oroville McCoy recalled his earlier years as a healthy, active outdoorsman, before Black Lung disease left him bedridden.

*Mingo County, West Virginia (1976)*

This older miner's hands show the brutal impact
of underground coal mining on the body.

*Price, Utah (1974)*

A coal mine roof fall victim pictured with his son
is but one example of the personal loss that
disabling injury brings to the victims and their
family members.

*Dingess, West Virginia (1973)*

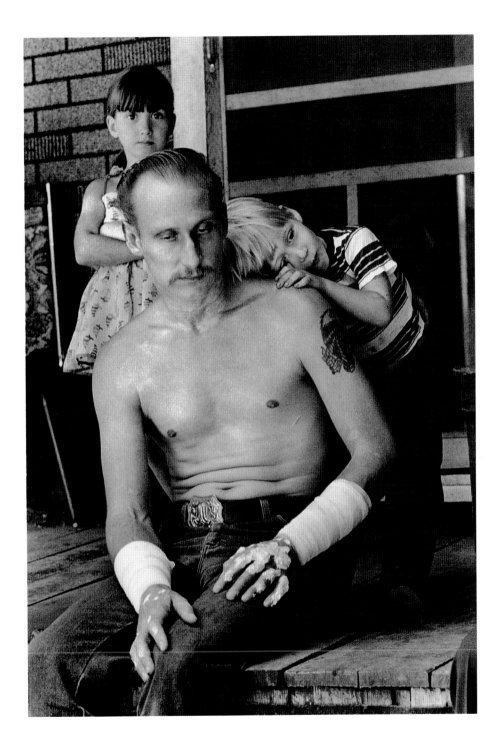

A survivor of a methane gas explosion at Pittston Coal Company's McClure mine recuperates from burns while at home with his family. The explosion claimed six lives.

*McClure, Virginia (1983)*

The wife of a coal mine rescue team member has just learned she is a widow. Her husband died in a second explosion with a team attempting rescue of 15 victims from a previous methane explosion in the Scotia Mine.

*Oven Fork, Kentucky (1976)*

Her husband survived Vietnam to die in a coal mine. The Scotia Mine Disaster claimed 26 lives in two explosions.

*Oven Fork, Kentucky (1976)*

The aftermath of the Scotia Mine Disaster.

*Oven Fork, Kentucky (1976)*

# AUTOMOBILE/HEAVY MANUFACTURING

The poor design of this electric wheel-welding device
showers the operator with hot metal particles.

*Romulus, Michigan (1977)*

Working in a wheel factory warehouse means this autoworker does a highly repetitive job in total isolation. Hefting heavy wheels onto the moving belt puts debilitating strain on all upper body parts.

*Romulus, Michigan (1977)*

An auto wheel maker stands among antiquated stamping presses with large, dangerously exposed moving parts. Modern presses of safer design are replacing the older machinery and can be seen in the background of the photograph.

*Romulus, Michigan (1977)*

An appliance assembler in a clothes washer factory handles a heavy impact wrench in each hand with relative ease. Using counterweights suspended from cables eliminates the dead weight of the tools, allowing the worker to work efficiently without risk to her musculoskeletal system.

*Newton, Iowa (1978)*

This autoworker has to overreach to install a roof rack on an automobile moving along the production line. Bad ergonomic design can cause disabling repetitive motion injury, poor workmanship, and reduced productivity.

*Detroit, Michigan (1976)*

Autoworkers perform assembly operations on a moving production line.

*Flint, Michigan (1977)*

An autoworker tests batteries as they move down the production line. In the storage battery industry, the absence of careful dust controls can result in lead poisoning. A lead-free workplace is particularly important for women of childbearing age.

*Visalia, California (1979)*

The asbestos fibers shaved off clutch plates are drawn into the filtered exhaust system and away from the autoworker. He is wearing an asbestos monitor on his shoulder to track the ongoing effectiveness of the controls.

*Detroit, Michigan (1978)*

This pottery worker dusts toilet molds with talc before pouring ceramic mix. The talc helps the unfired molded mix release easily from the mold. Exposure over time can result in respiratory disease known as talcosis.

*Plainfield, Connecticut (1986)*

Thousands of autoworkers have suffered because of downsizing of the U.S. automobile industry in the 1980s and '90s. For those left on the assembly line, the smaller work force has often meant compulsory overtime — a source of stress in any workplace. Above, an overworked autoworker takes a smoking break. *(1978)* Signs like the one at right have become an all-too-common sight.

*Detroit, Michigan  (1980)*

This safety sign speaks cryptically of the dangers steelworkers once endured at the Homestead Works. The steel mill stacks are all that remain of the Works that supplied jobs for thousands in the town for more than 100 years.

*Homestead, Pennsylvania (1992)*

# WOOD PRODUCTS

As this logger completes his final cut, he takes a preplanned exit route away from the cutting site. Logging remains one of the most dangerous jobs in the United States.

*Longview, Washington (1979)*

Loggers make a second chainsaw cut, enabling them to place hydraulic jacks for a controlled fall of a Douglas fir. Sustained use of chainsaws can cause nerve damage commonly known as "white finger."

*Longview, Washington (1979)*

An entire truckload of logs is hoisted off a truck bed as the driver
waits for the lumber mill log transport to move out of his way.

*Roseburg, Oregon (1992)*

As this logging accident attests, getting the logs successfully out of the
woods requires driver training, well-built logging roads, and constant vigilance.

*Wright City, Oklahoma (1981)*

This operator of plywood veneering machinery fixes a jam while the operator in background waits to restart the machinery.

*Longview, Washington (1979)*

This personalized machinery lockout tag is designed for someone to think twice before removing it.

*Williamsburg, Virginia (1995)*

The sawmill repairman's vulnerability is evident and demonstrates the need for a well-designed machinery lockout/tagout system.

*Montcure, North Carolina (1981)*

This furniture finisher is exposed to fumes from the petroleum-based varnish. His hands are also in direct contact with the wet varnish.

*Kenbridge, Virginia (1981)*

This injured lumber mill worker was paid to report to work, enabling the company to avoid a lost time injury report. By underreporting injury rate data, this wood products mill attempted to keep its workers' compensation insurance rates at an artificially low level.

*Franklin, Virginia (1984)*

# ACTIVISM

Cesar Chavez protests the use of pesticides in farm work. He brought attention to
the issue with a national campaign to educate the public about the dangers.

*Herald Square, New York City (1986)*

Left: Members of the Carolina Brown Lung Association demonstrate on the steps of the South Carolina statehouse.

*Columbia, South Carolina (1978)*

The CBLA was instrumental in securing enactment of OSHA's Cotton Dust Standard, in addition to obtaining substantial health and safety law improvements. The Roanoke Rapids chapter of the CBLA is shown above.

*Roanoke Rapids, North Carolina (1978)*

Unionized garment workers demonstrate against sweatshops
in Manhattan's Fashion District.

*Borough of Manhattan, New York (1995)*

Garment workers express their unity to sweatshop workers who are looking out windows as they work in Manhattan's Fashion District.

*Borough of Manhattan, New York (1995)*

Disabled miners and their widows confront their Kentucky congressman about
his poor record of supporting Black Lung reform legislation.

*Washington, D.C. (1975)*

Coal miners rally for Black Lung law reform on the steps of the U.S. Capitol.

*Washington, D.C. (1975)*

A hospital worker on strike asks the community to support her union's efforts in securing contractual workplace health and safety improvements.

*Prestonsburg, Kentucky (1981)*

Coal miners picket the mine entrance at dawn. Often, coal miners have to enforce their contractual right to refuse unsafe work orders.

*Sheridan, Wyoming (1974)*

# JOB SATISFACTION

Proud of her accomplishment, this autoworker expresses
the satisfaction of a rewarding, healthful, and safe job.

*Warren, Michigan (1982)*

A concrete finishing crew works in unison during the construction
of the Federal Triangle Office Building.

*Washington, D.C. (1994)*

These three coal miners were among the first women to work underground after
winning that right by taking their fight all the way to the U.S. Supreme Court.

*Vansant, Virginia (1976)*

A water quality technician tests a representative water sample from the city water supply.

*Chicago, Illinois (1993)*

A truck spring repairman poses with a tool of his trade.

*Detroit, Michigan (1978)*

A textile worker inspects nylon strands as he prepares to weave commercial belting.

*Buffalo, New York (1993)*

A wastewater treatment plant operator monitors a control valve.

*Honolulu, Hawaii (1984)*

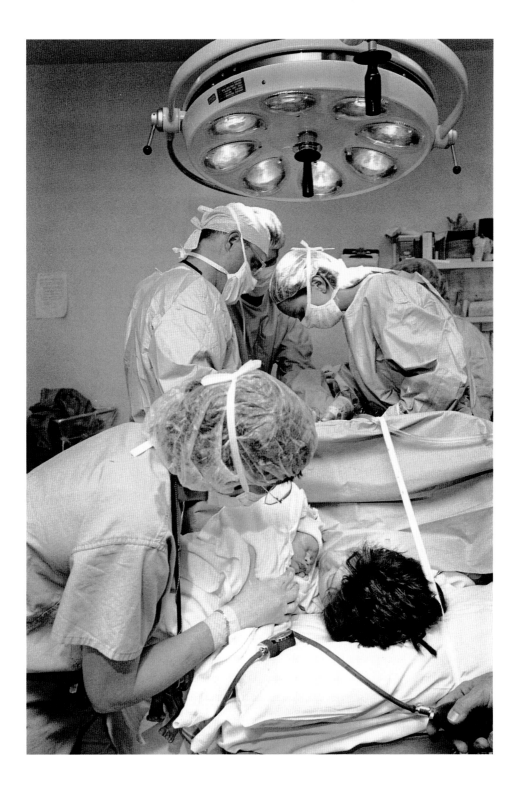

The labor of these nurses and physicians brought joy to a new mother after a difficult birth.

*Silver Spring, Maryland (1987)*

At right, the father of a newborn son savors the pleasures of parenting.

*Gaithersburg, Maryland (1990)*

An autoworker and son at breakfast.

*Buffalo, New York (1982)*

A long-haul truck driver enjoys the company of his family during his son's homework session.

*Appomattox, Virginia (1993)*

This steel mill town family is still hopeful as they take a Sunday afternoon stroll, despite the idle steel furnaces behind them.

*Braddock, Pennsylvania (1992)*

A child cools off in a summer shower.

*Alexandria, Virginia (1991)*

A school crossing guard with his students.

*Oak Park, Illinois (1993)*

In 1997 United Parcel Service employees gained widespread support from the public in their effort to reverse the long trend toward less satisfying part-time jobs.

*Washington, D.C. (1980)*

Helping hands pass the skills that only experience brings to a new generation of garment workers.

*Borough of Manhattan, New York (1980)*

After spending most of their working lives on the job together,
these autoworkers have forged an enduring bond of friendship.

*Detroit, Michigan (1978)*

A coal mine construction worker walks on high steel as the morning sun rises.

*Albers, Illinois (1976)*

# Afterword

## The Worker, the Photographer, and the Occupational Health Physician: Sharing Images, Sharing Goals

Howard Frumkin, MD, Dr PH, FACP, FACOEM
Chair, Department of Environmental and Occupational Health
Rollins School of Public Health of Emory University

The Civil War had Mathew Brady. America's national parks had Ansel Adams. In the early 20th century Lewis Hine was the one who provided society a mirror for the inequities of child labor. For the Depression, there were the likes of Walker Evans and Dorothea Lange. And for chronicling the trials and triumphs of the modern workplace, the name Earl Dotter can be added to this pantheon.

The allure of the photograph is enormous. Turn to the front page of any newspaper and you'll find it crammed with stories about Congressional hearings, budget deficits, and the ebb and flow on Wall Street. You will glance at the headlines, of course, but if you're like most people your eyes go straight to the photographs. A photograph can be so powerful, in fact, it eclipses the details of the story, swelling from image to icon, forever symbolizing an entire era. Who can forget the student kneeling over her fallen friend after the shootings at Kent State University; the smug, porcine Southern sheriff sitting defiantly at a civil rights trial; the Vietnamese child running toward the camera, burned with Napalm and frantic with pain.

In much this way, Earl's photographs endure. They capture the humanity, the nobility, and the struggles of working people; the pain of their workplace injuries and illnesses; the grief when their dreams are cut short, and the hope that sustains them. Anybody who works in occupational health and safety is familiar with the coal miner at shift's end: soiled, weary, but proud; the window washer dangling precariously from the Empire State Building; the 10-year-old girl picking cotton in a field in Mississippi. These are the visions that grace the pages of our textbooks, the walls of our offices and clinics, and the projection screens at the front of our classrooms. They are the powerful images that frame the way we see the working world.

Earl's photographs contribute much to the practice of occupational health and safety. Our profession calls for equal parts heart and head. We blend compassion and hard data. While we soar on our passion for social justice, we must be grounded in the discipline of scientific objectivity. One without the other is ineffective. Dr. Irving Selikoff is credited with saying, "Statistics are people with the tears wiped off." Statistics are easy: we collect them, analyze them, and talk about them so constantly they form a backdrop of all our work. But to balance the scales — reminding us of the tears — there is no better medium than the one Earl provides.

Because I teach occupational medicine to physicians, and occupational health to public health students, Earl's work has been especially useful to me. In these fields, as in many disciplines, we often define our teaching goals in terms of three dimensions: knowledge, attitudes, and behavior. If our teaching is successful, we achieve a measurable impact on one or more of these dimensions in our students. Earl's photographs are invaluable in supporting each of these occupational health teaching goals.

Earl's photographs increase our knowledge of occupational health and safety in obvious ways. Just as the angler must know all the varieties of fish, and the carpenter and construction worker must be expert in the use of their tools, occupational health students and professionals must recognize a wide range of production processes, job requirements, and associated hazards. Nothing achieves this knowledge as well as direct observation in the workplace, and field visits are an essential part of occupational health training. But accurate images are also invaluable. When I show Earl's photographs to my students, they can learn — quickly and in a structured manner — what the hazards look like, what health effects they cause, and how to prevent them.

Earl's images also help us shape students' attitudes. The photographer's art and the physician's practice, it seems to me, share important features. In both callings, there is an asymmetry — even a hierarchy — between the professional and the "client." In the doctor-patient relationship, the patient may be needy and vulnerable, while the doctor is the one with knowledge, power, and control. The skilled physician recognizes this inequity and takes steps to limit it, from fundamental commitments such as including the patient in treatment decisions to small gestures such as sitting to talk with a bedridden patient rather than towering over the bed.

The same tension exists between photographer and subject. The photographer, far more than the subject, determines how the image will be composed, cropped, and printed. The photographer has control. Here, as in the doctor-patient encounter, there is a danger of condescension, voyeurism, even exploitation. In this respect, Earl's work serves as a model not only for photographers but also for physicians. In every image, the subject's dignity is evident, mirroring Earl's deep respect for the people he photographs. It comes as no surprise that Earl knows the name of each of his subjects, a valuable example for doctors. In professional relationships, for both photographers and doctors, respect rather than condescension is the ideal.

This is a special challenge in occupational health because of class differences. The college-educated photographer is both artist and professional; the physician belongs to the archetypal elite profession. Both usually live a considerable social distance from the working class and poor people who labor at the nation's most dangerous jobs — the people Earl captures with his camera and those the occupational health professional sees as patients. To those who work at the boundaries, class differences are real and unavoidable. Doing our job well means remaining aware of class differences, withholding judgments, and patiently building alliances based on respect and shared ideals. Earl's photographs do not let us forget how important this attitude is.

Earl's work reminds us, and our students, of essential behaviors in occupational health. Go to the workplace. Make careful and accurate observations. Record them in detail so they can be used as the basis for conclusions. Throughout this process, exercise courtesy, respect, and compassion for patients.

Roman Vishniac, the renowned photographer and physician who recorded life in the Jewish settlements of Eastern Europe before the Second World War, wrote: "I was unable to save my people, only their memory." As the physician works to save people from death and disease, the photographer works to rescue people from oblivion. In *The Quiet Sickness*, the two missions fuse: we remember and honor the people to whom our professional efforts are dedicated.

The publication of this magnificent book is a landmark event in occupational health and safety. Many more occupational health physicians, nurses, industrial hygienists, related professionals, and students — not to mention working people, other photographers, and members of the public — will at last have the opportunity to gaze at these images and to learn from them. I am confident this work will help move us toward a goal Earl shares with members of my profession: better health and safety and dignity for all working people. His photographs deserve our deep admiration, and Earl our sincere gratitude.

# Autobiographical Notes

## Earl Dotter

Beginning in the Appalachian coalfields, and continuing through the last 25 years over a broad range of industries and regions of the country, I have observed and documented the working lives of Americans. Standing behind the lens, I have celebrated the accomplishments, the pride, and the skill of workers and community activists. In the course of my photography I have witnessed far too many workers of all descriptions experiencing a quiet sickness within their own bodies as the cumulative effect of long-term exposure to toxic dust and chemicals finally manifested itself. It wasn't long after I first observed the consequences of workplace fatalities and of lives shortened by disease that I found myself compelled to make a photographic record to humanize what has existed primarily as a statistical tragedy in the United States.

I began my photographic career after completing my studies at the School of Visual Arts in New York City. In 1968, I joined VISTA (Volunteers In Service To America) and was assigned to the Cumberland Plateau region of Tennessee. Over time, I was welcomed into the homes of coal-mining families. I came to know and respect their culture and struggles — a relationship that continues to this day.

After my VISTA assignment concluded, I remained in the area to photograph the rank-and-file movement to reform the United Mine Workers of America (UMWA), then under the corrupt leadership of Tony Boyle, who was later convicted and sentenced to life in prison. In 1972, I was invited to join the staff of the reformers' newspaper, *The Miner's Voice*, and subsequently became the photographer for the campaign to unseat Boyle, called the Miners for Democracy. When the election effort proved successful, I went to work for the *United Mine Workers Journal*, where I remained until 1977.

The emphasis of the *United Mine Workers Journal* was on improving miners' health and safety and the quality of life in their communities. My position enabled me to record the intimate aspects of daily life — the dangers of mining underground, the hardships of living on abused land, but also the joys, dignity, and culture that sustained coalfield families. It became a decade of intense creative development for me, during which I learned not just what to photograph, but how to create an image that would impact the viewer both visually and emotionally. The lessons learned during my "coalfield years" still guide my work today.

Throughout the 1980s, I photographed a wide array of occupational subjects. My photography has consistently been given life and texture by shooting not just the work, but the whole worker and his or her life on the job, at home, and in the community. Over the years, my subjects have expanded from an emphasis on occupational health and safety to include environmental hazards to public health. The evolution was only logical, since the adverse conditions that first affect people on the job as they take the "first hit" — from exposure to carcinogens, toxins, and industrial waste — eventually make their way out of the work site and into the air and water of the surrounding environment.

My photography was honored along with the staff of the *United Mine Workers Journal*, winning a National Magazine Award for Specialized Journalism in 1976. In 1988, I was pleased to receive the Leica Medal of Excellence for photographs featuring window washers working on the Empire State Building; to have many of my photos published in a definitive textbook titled *Occupational Health*, edited by Levy and Wegman (Little Brown & Company); and to take the photographs for a 26-page photo essay/article in *Audubon Magazine* titled "The Mountains, the Miners, and Mister Caudill."

In 1992, I completed the photos for the picture section of William Serrin's book *Homestead: The Glory and Tragedy of an American Steel Town* (Random House), depicting the death of small-town America as industry went through a dramatic decline attributable to technology and foreign relocation. The January/February 1997 issue of the *Columbia Journalism Review* features four pages of my photography in a photo/feature titled "Life's Work."

In the spring of 1996, I began a tour of my exhibit "The Quiet Sickness: A Photographic Chronicle of Hazardous Work in America," with the 120-picture exhibit featured at the American Industrial Hygiene Conference & Exposition in Washington, D.C. The exhibit (titled at the time "The Quiet Sickness: Occupational Hazards in the United States") was also shown during the 75th Anniversary Year Celebration of The Harvard School of Public Health in the spring of 1997.

My primary goal for the exhibit and this book is to put a human face on the occupationally related tragedies that befall thousands of U.S. workers annually. I strive to allow the subjects in my pictures to communicate directly with the viewer. One of the most gratifying responses from an industrial hygienist viewing the exhibit was his telling me that seeing the consequences of workplace-related injury, disease, and death faced by the real people who were featured reconnected him with his original motivation for entering the profession.

As a photographer, I am aware that the lives of my subjects are often far harsher and more painful than those who look at my pictures. To bridge that gap, I look for common ground the workers in the photos share with those who gaze upon them. Often, to be successful I need only capture my subject's desire for dignity and self-respect despite his or her pain and suffering. When I walk through a mine, mill, or factory, I find myself drawn to those subjects who emanate a sense of personal worth. When I experience tragedy in the workplace — death, disability, or exploitation — I use the camera to explore not just the person or event, but my own reaction to it. If I am successful, the viewer will be better able to stand before the photograph and feel the intensity of the moment as I felt it myself.

My goal is not just to touch those viewers already sympathetic to the circumstances of my subjects, but to command the attention of those who normally would pass them by.

Earl Dotter (on right), with Matt Witt, the former editor of the *United Mine Workers Journal*, while they prepared a profile of coal miners working in low coal seams.
*Logan County, West Virginia (1976)*